101 Fun Facts on Astronomy

By Kevin J Manning

1. A billion Suns can fit inside the largest star ever discovered, VY Canis Majoris.

2. The Andromeda Galaxy is the most distant object the human eye can see without optical aide.

3. Telescopes allow us to see the invisible!

4. The largest moon in the entire solar system, Ganymede, orbits the largest planet, Jupiter.

5.   Though the Earth seems suspended in space, it weighs in at some six thousand million million million tons.

6.   In one second, the speed of light will take us 7 ½ times around the Earth or to the Moon and back round trip in less than 3 seconds.

7.   Proxima Centauri is the name of the closest star to our solar system.

8.   The Milky Way Galaxy that we are a part of is on a collision course with the Andromeda Galaxy.

9.   Polaris is known as the North Star, but Thuban in Draco was our north star a few thousand years ago due to precession.

10.   A teaspoon of a neutron star's material would outweigh the Empire State Building in New York City.

11.   The supernova explosion witnessed in 1054 A.D. resulting in the Crab Nebula was so bright it was visible during the day and at night outshined the full Moon.

12.   Shooting stars are a misnomer as they actually are not stars at all but typically very small rocks and dust particles left behind by comets, known as meteors.

13.   Dark energy is believed to be the cause of the universe's accelerating expansion.

14.   Though Pluto is no longer considered a planet by modern definition, it still lives up to the original

meaning of the word as "wanderer" and has 5 moons orbiting it.

15.   Though it is further out from the sun, the planet Venus is hotter than Mercury due to its dense atmosphere trapping the gases.

16.   The inhabitable zone for an extrasolar planet is the distance from its parent star where water could exist in liquid form.

17.   Comets are essentially "dirty snowballs" travelling from cold distant regions in space known as the Kuiper Belt and Oort Cloud.

18.   As much water exists below on the Earth, a great deal of water also exists above the clouds. Evidence tells us there is water on the Moon, Mars, asteroids, comets, Jupiter's moon Europa, Saturn's

moons Titan and Enceladus, protoplanetary disks, and extrasolar planets.

19. Seasonal temperature changes are caused by 2 things: the inclination of the Sun's rays and the length of day and night.

20. Meteors create a streak of light across the sky from ram pressure causing excitation of the atoms.

21. Most of the nearest stars to our solar system are so faint they are invisible to the unaided eye.

22. A light year is the distance light travels in an entire year, which is a little less than six trillion miles.

23. Right ascension and declination are the celestial coordinates representing

latitude and longitude projected up in the sky. If you face north, east is to your right, where celestial objects appear to be ascending from the Earth's rotation.

24.   Essentially, the only two types of optical telescopes are refractors and reflectors.

25.   One parsec equals the distance of 3.26 light years, and the name came from the parallax shift of one arc second using the orbit of the Earth about the Sun as the baseline.

26.   Modern astronomy uses detectors sensitive in all major regions of the electromagnetic spectrum so more information can be collected from a given object.

27.   The word nebula comes from the Greek meaning "cloud," and is a cloud of gas and dust in space.

28.   The brightest star appearing in the night sky is named Sirius in the constellation Canis Major visible early in the winter sky.

29.   Sir Isaac Newton created the optical design of the classical Newtonian reflector telescope.

30.   The same piece of rock in space is given 3 different names depending upon its position – a meteoroid in space, a meteor as a streak of light across the sky, and a meteorite that has landed on the ground.

31.   Each year the Moon is moving further away from the Earth by a little more than an inch.

32. Since its discovery in 1846, the planet Neptune will have orbited the Sun once on July 12, 2011.

33. Since 1930, there are 88 official constellations dividing the entire night sky north and south.

34. Saturn's density is so low, if you could place it in the ocean it would float.

35. The 2 moons orbiting the planet Mars, Phobos and Deimos, are considered to be captured asteroids.

36. Astronomy is the oldest of all the sciences, and lends itself readily to multidisciplinary connections.

37. Approximately 1 undecillion (10^{36}) atoms could stretch across the known observable universe.

38.　The core of the Sun is about 27 million degrees Fahrenheit.

39.　An astronomical unit (AU) is the distance from the Earth to the Sun, averaging about 93 million miles.

40.　Comet Shoemaker-Levy 9 was a comet that broke apart into 21 fragments after coming within Jupiter's Roche limit and collided with Jupiter in July 1994, an unprecedented event that provided the first observation of direct impacts by large bodies on another planet.

41.　Although you may feel like you are sitting completely still in your chair, we are all constantly moving six different ways, all at the same time, each way violently fast.

42.   Gravitational interaction with the Moon is causing the Earth's rotation to slow down a very small amount, increasing the length of a day.

43.   Due to the solar wind, the tail of a comet always points away from the Sun.

44.   The Earth rotates faster at the equator than it does at the poles.

45.   It would take 100,000 years to cross the Milky Way Galaxy at the speed of light.

46.   Lunar libration is a slight wobbling or rocking of the Moon allowing us to see more than half, nearly 59% of the lunar landscape.

47.   The 4 Jovian worlds, or gas giants, Jupiter, Saturn, Uranus, and Neptune all have rings.

48. Only about 5% of the entire universe is detectable and observable. The vast majority of it is the strange dark energy, with the rest being dark matter.

49. On the Moon, you would weigh only one sixth as much as you do here on Earth.

50. An annular solar eclipse occurs when the Moon is further away in its orbit and isn't large enough to completely cover the Sun, leaving an annulus ring of fire.

51. Virtual images as seen in telescopes are normally inverted.

52. The "power" of a telescope is its ability to gather and focus light from the area across the primary objective, so the bigger the better.

53.   All natural celestial objects appear to move across the sky from east to west solely because of the Earth's rotation each day.

54.   Galaxies typically form in clusters and superclusters of stupendous size.

55.   A gravitational lens is formed by the huge gravity combined within a cluster of galaxies allowing us to "see" a much more distant galaxy beyond.

56.   High energy electrically charged particles held in place by the Earth's magnetic field is known as the Van Allen radiation belts.

57.   An aurora, commonly called the northern or southern lights, is produced when electrically charged particles from the solar wind interact with the Earth's upper atmosphere and fluoresce.

58. What came first, the supermassive black hole or the galaxy? Coevolution is a theory that presumes that both evolved together as evidenced by the interaction between the two.

59. There are 110 Messier Objects, whose discoverer Charles Messier at first thought these galaxies, nebulae and star clusters resembled a comet in deep space.

60. Ultraviolet rays travelling from nearby stars push clouds of gas and dust into formations resembling bubbles in space.

61. The Trifid Nebula's name means 'divided into three lobes,' containing three different types of nebulosity in the form of an emission nebula, a reflection nebula and a dark nebula.

62.   Some open (galactic) star clusters appear as diamonds set in black velvet, containing tens to hundreds of hot young stars.

63.   To calculate magnification for any telescope, divide the focal length of the telescope by the focal length of the eyepiece, or ocular, using the same units.

64.   The numbers marked on the backs of binoculars, such as 7x35, means that objects through them will appear 7 times larger in diameter than with the unaided eyes and the front lenses are 35mm in diameter.

65.   The Large Binocular Telescope (LBT) is the world's largest optical telescope on a single mount, having two 27.6 foot mirrors that collectively will

resolve stars equivalent to a single mirror 74.8 feet across.

66.   Most stars in our Milky Way Galaxy are obscured by gas and dust, so the stars bright enough to be seen by the average person on a dark clear night number about 3,000.

67.   The colors of stars infer something of their temperature, with red being the coolest and blue being the hottest stars, just the opposite of what an artist would use on a painting.

68.   Like chandeliers surrounding the core of the galaxy in a region called the halo, globular star clusters are roughly spherical regions containing thousands to over a million old stars.

69.   Ejnar Hertzsprung and Henry Norris Russell independently came up with a

correlation amongst the stars that has literally revolutionized our understanding of stars and stellar life cycles, known as the Hertzsprung–Russell (HR) diagram.

70. The Earth's atmosphere, with its water vapor, dust particles, and turbulent air currents, hinders our view of the universe, so space telescopes have been designed to eliminate that problem.

71. The word neutrino literally means "little neutral particle." Since neutrinos are subatomic particles that are produced by reactions in the Sun's core, are able to pass through the outer layers and continue their journey to the earth with very weak interactions with matter, many billions of these particles pass through our body, rocks, even lead each second.

72. A 1st magnitude star is 100 times as bright as a 6th magnitude star.

73. The point directly overhead from an observer is known as the zenith, while the point directly below is called the nadir.

74. The apparent shift of the celestial poles caused by a gradual wobble of the Earth's axis that lasts nearly 26,000 years is referred to as Precession of the Equinoxes.

75. An event that occurs when one object is hidden by another object that passes between it and the observer, such as the Moon passing in front of Jupiter in our line of sight, is called an occultation.

76. A total lunar eclipse can only occur during a Full Moon, and a total solar

eclipse can only occur at the New Moon phase.

77.   A large, brilliant meteor brighter than the planet Venus is called a Fireball. When this rock explodes with a thunderous sound from trapped gases then it is called a Bolide.

78.   A Fall is when you see a meteor land on the ground as a meteorite, and a Find is when you stumble on one without having seen it land.

79.   The glow of the Milky Way stretching across the summer sky is literally thousands of unresolved stars. Even a small pair of binoculars will resolve the glow into separate points of light.

80.   Diamonds are raining down from space upon the Earth in the form of

nanodiamond dust contained in certain meteorites.

81. The Big Dipper acts as a signpost in the sky for locating two different stars. The two stars at the end of the bowl opposite the handle are known as "pointers" as they point in the direction of Polaris, the North Star. The handle is said to "arc" toward Arcturus in the constellation Boötes.

82. It's been said that the Sun is 400 times further away from us than the Moon, and that it is also 400 times larger in physical diameter than the Moon as well. So the size and distance ratios are precisely identical upon average, which is why we experience a total solar eclipse as we do, and if not, would have been major consequences for Albert Einstein.

83.   A Super Full Moon is when the Moon is at Perigee in its orbit, making it closer than at other points, at the same time it reaches the Full Moon phase, which occurs upon average every 18 years or so.

84.   There are birds that migrate by recognizing patterns of stars in the sky.

85.   No two stars are likely to contact each other when galaxies collide in spite of the hundreds of billions of stars involved. So vast is the space between them.

86.   The remnant energy left over from the Big Bang accounts for why the coldest areas thought to exist in the universe is at a temperature of about 3 Kelvins, and is known as the Cosmic Microwave Background (CMB) radiation.

87. Terahertz waves are now being studied as part of the electromagnetic spectrum and are found to have interesting properties, and lie just on the border of the microwave and infrared waves.

88. Dark matter is called dark because it does not interact with any form of electromagnetic radiation, yet gravitational effects on visible matter in the universe gives evidence to its existence.

89. 16 of the 18 types of elementary particles predicted by quantum physics have been detected by experiment already, with the goal of searching for the remaining two particles, such as the Higgs Boson and the Graviton.

90.   Solar neutrinos go through 3 oscillations en route from the Sun, from an Electron Neutrino to a Muon Neutrino to a Tau Neutrino, which is why only one third the number expected were detected initially.

91.   Venus, Uranus, and Pluto have retrograde rotations about their axes, meaning that these bodies spin in the opposite direction than their orbits.

92.   A high degree of differentiation among the planets during the early formation of the solar system occurred due to a temperature gradient, which determined how material condensed according to their distance from the Sun.

93.   Mathematically, White holes are a perfectly valid solution to the equations of general relativity because they are symmetric in time, but they most likely

do not exist in nature. For the same reason, wormholes almost certainly do not exist either, and if they did they would not be stable and you would get fried by X-rays and gamma rays.

94. As Star Trek reminded us, we are carbon-based life forms, but the Big Bang didn't actually produce any carbon. The excited state of the carbon-12 nucleus, called the Hoyle state, is necessary for the fusion of three alpha particles in the interior of stars.

95. In our quest to look for other worlds like our Earth, the Kepler Mission uses the transit Method of detecting extrasolar planets that pass in front of their parent star in our line of sight and thereby cause a small, but measurable change in a star's brightness. This must repeat the same each time periodically; then we can calculate the planet's orbital

size, the mass of the star, the sizes of the planet and the star. From the temperature of the star, we can infer if the planet is within the habitable zone of a wide variety of stars.

96.   Due to the rotation of the Earth, the Moon will appear to move across the sky an amount equal to its own diameter every two minutes.

97.   The smallest blur spot that a lens can produce at a focus is referred to as the least circle of confusion.

98.   Jupiter's Great Red Spot is a cyclonic storm larger than the Earth and has been going strong for at least 400 years since Galileo first identified it in his telescope.

99.   The diagonal or secondary mirror in a reflecting telescope creates a real

central obstruction to the light path, but the diffraction and light loss is negligible if kept to near 10% of the aperture of the primary mirror.

100.   Gamma-ray bursts (GRBs) are flashes of gamma rays associated with extremely
   energetic explosions in distant galaxies, and are considered the most luminous electromagnetic events occurring in the universe.

101.   When the most massive stars explode, they are referred to as a hypernova.

Kevin Manning is an international award winning astronomer, and while he is supposed to be retired, he's as busy as ever doing a nationwide Star Tour! Kevin won national and international awards in his field, and did some work with Brookhaven National Laboratory. Kevin was both a Wright Fellow at Tufts University and an Einstein Fellow working on Capitol Hill in Washington, DC. He served as an Editor for the U.S. Department of Energy's Office of Science Journal of Undergraduate Research, 2001 (vol. 2), and worked with the International Atomic Energy Agency (IAEA) with the U.S. Support Program (USSP). Besides the numerous workshops he's presented over the years at libraries, observatories, and science centers, some noteworthy ones include those made at Tufts University, State University of New York at Stony Brook, the NSTA National Convention, AAAS Breakfast with Scientists, and the National Parks Service. While Kevin's scientific background is stellar and his mind is definitely in the stars, he's also one of the most down to earth and accessible guys you'll ever have the chance to meet. Kevin goes to great lengths to communicate his passion for the amazing world of outer space in terms that all ages can understand. He's lectured widely, built telescopes and observatories, and consulted with NASA, but his favorite thing to do is to share his contagious enthusiasm and knowledge with the public. Astronomy is more than a study for Kevin, it's his love and it shows in all he does.

www.ingramcontent.com/pod-product-compliance
Lightning Source LLC
Chambersburg PA
CBHW070800180526
45168CB00004B/1687